Echo 2nd Generation Manual

Echo 2nd Generation User Guide

By Emery H. Maxwell

Disclaimer:

The views expressed within this book are those of the author alone. *Echo* and *ALEXA* are trademarks of *Amazon*. All other copyrights and trademarks are properties of their respective owners. The information contained within this book is based on the opinions, observations, and experiences of the author and is provided "AS-IS". No warranties of any kind are made. Neither the author nor publisher are engaged in rendering professional services of any kind. Neither the author nor publisher will assume responsibility or liability for any loss or damage related directly or indirectly to the information contained within this book.

The author has attempted to be as accurate as possible with the information contained within this book. Neither the author nor publisher will assume responsibility or liability for any errors, omissions, inconsistencies, or inaccuracies.

Table of Contents

Welcome

Welcome to the *Echo 2nd Generation Manual*. This user guide is intended to help you understand, set up, and manage *Amazon Echo*.

Using the wake word *ALEXA*, the device will hear you from across the room and will respond to your requests accordingly.

This guide will cover:

• Ring light colors and their meanings

• How to set up *Echo*

• How to teach *ALEXA* your voice

• How to connect to *BLUETOOTH*

• How to connect to external speakers

• How to play music on multiple *Echo* devices

• How to enable *skills*

• How to make and receive calls between compatible devices

• How to connect *Smart Home* devices to *ALEXA*

• How to control multiple *Smart* devices simultaneously

• How to use *Echo* with *IFTTT*

• Troubleshooting

• . . . and more.

It's time to get started.

Getting Started

To help you understand *Echo*, this section will cover the specifications of *Echo*, ring light colors and their meanings, and how to set up the device.

Specifications

Size: 5.8" x 3.4" x 3.4"

Weight: 29 oz. (Actual size and weight may vary by manufacturing process)

Audio: 2.5 " woofer and 0.6" tweeter. Can also connect to external speakers, 3.5mm stereo audio output. (Audio cable is not included).

WI-FI Connectivity: Dual-band WI-FI supports 802.11 a/b/g/n (2.4 and 5GHZ) networks. Does not support connecting to peer-to-peer WI-FI networks.

BLUETOOTH Connectivity: Advanced Audio Distribution Profile (A2DP) support for audio streaming from your mobile device to *Echo Dot* or from *Echo Dot* to a *BLUETOOTH* speaker. AVRCP for voice control of connected mobile devices. Hands-free voice control is not supported for Mac OS X devices. *BLUETOOTH* speakers requiring PIN codes are not supported.

External Display (Basic Hardware)

Volume Up: Button with the Plus sign.

Volume Down: Button with the Minus sign.

Microphone Off: Pressing this button will power off the microphones. Pressing it again will power the microphones back on.

Action: Button with the dot. Pressing this button will power off a timer or alarm or wake the device. Press and hold this button until the light ring changes to orange to enable WI-FI setup mode.

Audio Output: An audio cable can be used to connect *Echo 2nd Generation* to external speakers or home entertainment systems. However, audio cannot be played on *Echo 2nd Generation* from other devices.

Micro-USB Power: Slot next to the audio output.

Light Ring: Let's you know what the status is for the device.

Ring Light Colors and Their Meanings

- **Lights off:** The active device is waiting for your request.

- **Blue with spinning cyan lights:** The device is powering up (starting).

- **Blue with cyan pointing in direction of person talking:** ALEXA is processing the request.

- **Spinning blue light that ends in a purple light:** Do Not Disturb has been enabled.

- **White light:** Person is adjusting volume level on the device.

- **Solid red light:** Person has turned off the microphones on the device. The microphones can be powered back on by pressing the *Microphone* button.

- **Orange light spinning in a clockwise motion:** The device is connecting to your WI-FI network.

- **Continuous fluctuating purplish light:** An error has occurred during the WI-FI setup process.

- **Yellow light:** A message is waiting. To listen to the message, say, "Play my messages."

- **Green light:** Person is receiving a call or Drop In on the device.

- **Single flash of purple light**: Do Not Disturb is active.

Setup

Download the *ALEXA* app to your computer, tablet, or smart phone. On your mobile device, the app is available at *Apple App Store, Google Play,* and *Amazon App Store.*

1.) Go to one of these app stores on your mobile device and search for "*ALEXA app.*"

In order for the *ALEXA* app to work on 1st generation *Echo*, your phone or tablet must be:
- *Fire OS 3.0* or higher
- *IOS 8.0* or higher
- *Android 4.4* or higher

In order for the *ALEXA* app to work on 2nd generation *Echo*, your phone or tablet must be:
- *Fire OS 3.0* or higher
- *IOS 9.0* or higher
- *Android 5.0* or higher

On your WI-FI enabled computer, the *ALEXA* app can be downloaded on *Amazon's* website. If you are using Internet Explorer, it must be Internet Explorer 10 or higher.

2.) After the *ALEXA* app is downloaded, sign in. The sign-in information should be the same as your regular *Amazon* account sign-in information.

3.) There should be a power cord that came along with your *Echo* device. Connect the cord to the port on the device and plug the other end into an outlet.

The light ring at the top of the *Echo* device will turn blue, then orange. The voice will greet you and let you know that the *Amazon Echo* device is ready for setup.

Note: If you wait too long, the light may turn blueish purple. If that happens, press and hold the action button until the light turns back to orange. Then open the *ALEXA* app and go to **Settings > Set up a new device**.

4.) Now it's time to set up a WI-FI connection.

Set up WI-FI

- In the *ALEXA* app, go to **Settings**. The **Settings** section can be found in the left navigation panel.

- Select your device, then select **Set up a new device**.

• If you are using a 1st generation *Echo,* or if the set up process doesn't begin automatically, press and hold the *action* button until the light ring changes to orange. Then go to **Settings > Set up a new device.** You should now see a list of WI-FI networks in the app.

• Follow the instructions in the *ALEXA* app to connect the *Echo Dot* device to a network.

If *Echo* doesn't connect to your WI-FI network, unplug the device and plug it back in again. This will restart it.

• Select your WI-FI network and enter the password if prompted. If you don't see your WI-FI network, scroll down and select **Add a Network**. The name of your network can be found on your router. You can also select **RESCAN**, then wait and see if your WI-FI network comes up.

• When set up is complete, you should be able to talk to *ALEXA*.

How to Teach *ALEXA* Your Voice

Teaching *ALEXA* to recognize your voice can help create a more personalized experience across certain supported features.

This can be achieved by creating a voice profile.

Create a Voice Profile

1.) On your *smart phone*, go to the menu and select *Settings*.

2.) Go to the *Accounts* section.

3.) Select **Your Voice**.

4.) Select **Begin**.

5.) Using the drop-down menu, select the device you want to interact with to teach *ALEXA* your voice.

6.) Select **Next**.

7.) When prompted, say the phrase out loud. Then select **Next** to go to the next phrase. You can also try the phrase again by selecting **Try Again**.

8.) Select **Complete**.

You should now see a confirmation page on the screen.

To confirm, ask *ALEXA*, "Who am I?" If the process went well, *ALEXA* will mention your name.

How to Connect Your Device to BLUETOOTH and Audio Cable

T he following chapters are intended to help you connect *Echo* to *BLUETOOTH* and *Audio Cable*.

How to Connect to BLUETOOTH

On the *Select how you want to use your Echo* page, you will have the option to select *BLUETOOTH, Audio Cable,* or *No speakers.*

Before you begin, make sure you are using certified speakers that support *BLUETOOTH* profiles for *Echo* devices.

Supported BLUETOOTH Profiles:

• Advanced Audio Distribution Profile

• Audio / Video Control Profile

When you're ready to get started, follow the steps below.

1.) Power on the BLUETOOTH speaker and make sure the volume is not turned down too far.

2.) Since *Echo* can only connect to one BLUETOOTH device at a time, make sure other BLUETOOTH devices are disconnected from the *Echo* device.

3.) On the *BLUETOOTH* speaker, make sure *Pairing Mode* is on. If you are not sure how to enable or disable *Pairing mode*, check the *BLUETOOTH* speaker's owner's manual.

4.) In the *ALEXA* app, select *Settings*.

5.) Select your device, then select **BLUETOOTH**. Select **Pair a New Device.**

Your *Echo* device should now enter pairing mode.

The speaker will appear in the list of available devices in the *ALEXA* app when *Echo* finds and recognizes your *BLUETOOTH* speaker.

6.) Select your *BLUETOOTH* speaker.

Your *Echo* device will now connect to the speaker.

7.) To finish pairing your *Echo* device with your BLUETOOTH speaker, select *Continue* in the *ALEXA* app.

Once the device is paired, tell *ALEXA* to "Connect." The *Echo* device will connect to the device it was most recently paired with.

How to Connect to External Speakers with an Audio Cable

Although the *Echo* device already has a speaker, you might like to connect it to a more powerful one.

The 3.5mm audio cable is not included in the box of the *Echo* device, so it must be purchased separately.

1.) Locate and the AUX OUT on the *Echo* device. It should be next to the USB slot.

2.) Power on the external speaker.

3.) Plug one end of the cable into the *Echo* device and the other into the speaker.

Try to keep the external speaker at least a meter away from the device.

How to Play Music on Multiple *Echo* Devices

When enabled, *Multi-Room Music* allows you to control and play music through multiple *Echo* devices simultaneously.

Note: In order to stream different music on multiple devices or on multiple *Multi-Room Music Groups* simultaneously, you need to have an *Amazon Music Unlimited Family plan*. Under this plan, each family member (up to six in total) have their own account. In their accounts, they will have the ability to build their own music library, have unlimited streaming on mobile devices, enjoy personalized recommendations, and have the ability to make music available for offline playback.

Compatible Content

- *Amazon Music*
- *Amazon Music Unlimited (Family plan or Individual plan)*
- *Prime Music*
- *SPOTIFY*
- Certain third-party music providers, such as *TUNEIN*

Compatible Devices

- *Amazon Echo (1st and 2nd Generation)*
- *Echo Dot (1st and 2nd Generation)*
- *Echo Show*
- *Echo Plus*

Create a Multi-Room Music group in the ALEXA app

1.) Go to the menu, then select *Smart Home*.

2.) Select *Groups*.

3.) Select *Create Groups*.

4.) Select *Multi-Room Music Group*.

5.) From the drop-down menu, use existing group names or create your own with *Custom Name*.

6.) Choose which devices to include, then select *Create Group*.

Note: *Multi-Room music* will disable all *BLUETOOTH* connections while it's being used, including *BLUETOOTH* enabled *ALEXA* voice remotes and external speakers.

How to Enable Skills

Skills are voice-controlled capabilities that improve the *ALEXA* device's functionality. To illustrate, if you'd like *ALEXA* to tell you about specific upcoming events in your city, you would need to enable a specific skill for that.

Oftentimes, if you know the specific name of the skill you'd like to use, you can simply say, "*ALEXA*, enable [skill name]."

But sometimes certain skills need to be enabled through the *Amazon* website or the *ALEXA* app, while others might need to be activated by following the prompts from *ALEXA*.

Enable *Skills*

1.) Open the *ALEXA* app.

2.) Go to the menu and select **Skills**.

You can also go to the *Amazon* website and go into the *skills* section.

3.) Use the *search bar* to find a specific skill or browse through the skills by category.

4.) After finding the skill you'd like to use, select it to go to its detail page. The detail page should include at least one example of what to say to play or open the skill.

5.) On the skill's detail page, select **Enable Skill**.

Now you should be able to tell *ALEXA* to open the skill.

If you need help with the skill, say, "ALEXA, [skill name] help."

Manage *Skills*

1.) Open the *ALEXA* app.

2.) Go to the menu and select **Skills**.

3.) Select **Your Skills**.

4.) Select a *skill* to go to its detail page.

You should now see a list of available options.

<u>*ALEXA* Calling and Messaging</u>

W ith *ALEXA Calling and Messaging*, you can make and receive calls and messages between compatible *Echo* devices or the *ALEXA* app.

Most mobile and landline numbers in North America can be called, and it is available to anyone who has access to your compatible devices.

Anyone who has your information and chooses to use the feature can contact you on your compatible *Echo* device or *ALEXA* app.

It is a free service, but signing up is required.

<u>Compatible Devices</u>

- *ALEXA* app

- *Amazon Echo* (1st and 2nd Generation)

- *Echo Dot* (1st and 2nd Generation)

- *Echo Show*

- *Echo Spot*

- *Echo Plus*

How to Sign Up

In order to sign up for *ALEXA Calling and Messaging*, you will need:

• An *Amazon* account

• The *ALEXA* app installed on a compatible *Android* (OS 5.0 or higher) or *IOS* phone (9.0 or higher)

Currently, *ALEXA Calling and Messaging* is only available on *smart* phones. Tablets will not work with this service.

Sign up and set up the calling and messaging feature

1.) Open the *ALEXA* app on your compatible *IOS* or *Android* phone.

2.) Open the *Conversations* tab at the bottom of the menu.

3.) Following the instructions, enter you mobile phone information.

If done correctly, it will import your list of contacts for you.

You might be prompted to verify your mobile number through a text message.

Although *ALEXA Calling and Messaging* is considered separate from your phone service, the *ALEXA* app might use data if your phone is connected to the internet through your mobile network.

How to Use *Echo* to Make Calls

T o make a call using the *Echo* device, simply ask *ALEXA* to call the person you want to reach, mentioning the contact's name.

To make a call to the *ALEXA* app or another *Echo* device, say:

"ALEXA, call [person's name] Echo."

To make a call to a landline or mobile number that is saved to your list of contacts, say:

"ALEXA, call [person's name] on his/her home phone."

"ALEXA, call [person's name] mobile."

"ALEXA, call [person's name] at work."

"ALEXA, call [person's name] office."

To verbally dial a mobile or landline without saying the person's name: (Available on *Echo* devices only)

"ALEXA, call [say each digit, including the area code]."

To control the volume, say:

"ALEXA, turn the volume up / down."

You can also mute the line manually by using the **Microphone off** button on the device.

To hang up, say:

"Hang up."

"End call."

If you are making the call from the *ALEXA* app, you can also select the on-screen **End** tab to disconnect.

At this time, *ALEXA* does not support calls to certain types of numbers, such as:

• Emergency services

• Premium-rate or toll numbers

• International numbers outside of North America

• Dial-by-letter numbers

• Abbreviated dial codes

How to Make Calls in The *ALEXA* app

T here is also the option to call your *ALEXA – to – ALEXA* contacts from the *ALEXA* app.

1.) Open the **Conversations** tab at the bottom of the menu.

2.) Select the **Contacts** icon,

3.) You should now be able to view your list of contacts. Select the contact you would like to call.

4.) Select the *phone* icon to place the call.

To end a call from the *ALEXA* app, select the **End** button.

How to Answer or Ignore Calls

When a call comes in, a green light will flash on the supported *Echo* device. *ALEXA* will also alert you, announcing the caller's contact name.

ALEXA might ask you if you'd like to answer the call. But you can also say, "Answer," or "Ignore."

The green lighting will remain on the device for as long as the call is connected.

If another call comes in while you are still on the line with a different caller, the new call is automatically sent to another device.

If you'd like to temporarily block alerts for calls and messages on the *Echo* device, use the **Do Not Disturb** feature, and say, "Do not disturb me." To switch off the feature, say, "Turn off Do Not Disturb."

The **Do Not Disturb** feature can also be scheduled.

To schedule **Do Not Disturb** in the *ALEXA* app:

1.) Select *Settings* from the menu.

2.) Select your device.

3.) Look under **Do Not Disturb**, then select **Scheduled**.

4.) Using the toggle switch, enable the feature.

5.) Select **Edit**.

6.) You should now be able to edit the start and stop times. Select **Save Changes** when you're done.

How to Use *ALEXA* Messaging

Although *ALEXA* can't send photos or other attachments, it can deliver messages to the recipient's *ALEXA* app and other supported *Echo* devices.

Send Messages in The ALEXA app

1.) Select the *Conversations* icon.

2.) Select the *New Conversations* tab.

3.) You should now be able to see your list of contacts. If you'd like to start a new message, select a contact from the list. You can also continue an existing conversation by selecting the conversation that appears.

4.) Select the text tab to open the keyboard.

5.) When you're done typing the message, select the *Send* tab.

Check Messages in The ALEXA app

1.) Tap the *Conversations* icon.

2.) Select the new message with the green dot.

How to Send and Listen to Voice Messages with The *ALEXA* App

V oice messages can be sent and received through the ALEXA app.

The recipient's *Echo* device will make a chiming sound and turn yellow when a message is received.

Send Voice Messages From ALEXA App

1.) Select the *Conversations* icon at the bottom of the menu.

2.) Select the *New Conversations* tab and select a contact from the list. You can also respond to an existing conversation by selecting from the list of conversations shown.

3.) Press and hold the *Microphone* tab.

4.) Continuing to hold the *Microphone* tab, record your message.

5.) When you're done recording the message, release the *Microphone* tab. The message should now be sent.

Listen to Voice Message in The ALEXA app

1.) Select the *Conversations* icon at the bottom of the menu.

2.) Select the message.

3.) Press the *Play* icon. You can also read the message if it's transcribed into text.

How to Send and Listen to Voice Messages with The *Echo* Device

V oice messages can also be sent and listened to through a supported *Echo* device.

Send Voice Message Using *Echo*

1.) Say, "Send a message to [name of contact]."

2.) For confirmation, *ALEXA* might repeat the name back to you.

3.) Confirm the contact's name.

4.) *ALEXA* should now prompt you for the message.

5.) *ALEXA* will send your message when you're done talking.

Play Voice Messages on *Echo* Device

The recipient's *Echo* device will make a chiming sound and turn yellow when a message is received.

To have the messages played out loud, say, "Play my messages."

If there aren't any new messages, *ALEXA* will ask you if you'd like to hear the old ones.

For multiple household members, say, "Play messages for [household member's name]."

How to Manage Your Contacts and Settings

Managing your *ALEXA Calling and Messaging* settings and *ALEXA – to – ALEXA* contacts can be done from the *ALEXA* app.

How to Manage Your Profile

1.) Tap the *Conversations* tab at the bottom of the menu.

2.) Tap the *Contacts* icon.

3.) Select *My Profile.*

There should now be a list of options, including **Name, Drop In,** and **Caller ID**.

• **Name:** Allows you to change your profile name.

• **Drop In:** Enables you to allow or refuse permission to Drop In for a specific contact. Use the toggle switch to enable or disable this feature.

• **Caller ID:** Allows you to select whether or not you want ALEXA to display your contact information when making calls. Use the toggle switch to enable or disable this feature.

How to Manage Your Contacts

Although the *ALEXA* app does not display all of the contacts saved to your phone, it does display your *ALEXA – to – ALEXA* contacts.

Depending on the type of phone you have, go to *Contacts* or *Address Book*. From there, you should be able to add a contact or edit existing ones.

How to Block an *ALEXA – to – ALEXA* Contact

1.) Open the *ALEXA* app.

2.) Tap the *Conversations* icon at the bottom of the menu.

3.) Scroll down and tap **Block Contacts**.

4.) Select the contact you would like to block.

5.) Select **Block**, then confirm.

How to Set Up Profiles For Multiple Household Members

1.) Tap the *Conversations* icon at the bottom of the menu.

2.) Select **Get Started**.

3.) Add and verify your mobile number.

4.) When you reach the **Help ALEXA get to know you** page, you will have the option to select a profile you have already saved. If you do not have a profile saved, select **I'm someone else**.

5.) Enter your name and mobile phone information.

Each additional user can follow these same steps.

How to Connect *Smart* Home Devices to *ALEXA*

Before connecting a *smart home* device to *ALEXA*, it's important to be familiar with the safety information.

Safety Guidelines

• Follow the instructions for *smart home* devices.

• After a request is made, confirm the action has been completed on the *smart home* device.

• Make sure your *ALEXA* supported device and connected products are running efficiently. For example, make sure the *lock doors* feature is working before you leave your home.

How to Connect a *Smart Home* Device to *ALEXA* in The *ALEXA* App

1.) Go to the menu and select **Skills**.

2.) Search for and locate the skill you are looking for, then select **Enable**.

3.) Follow the on-screen directions to get through the linking process.

4.) Tell *ALEXA* to discover the device by saying "Discover my devices." You can also go into the **Smart Home** section in the *ALEXA* app and select **Add Device**.

How to Discover *Smart Home* Devices without a skill

Not all devices require a skill to connect to *ALEXA*.

If you'd like to connect these devices, simply tell *ALEXA* to "Discover devices."

Some devices might need to be powered on before they can be discovered. For example, if you are using a *Phillips Hue Bridge*, press the button on the bridge before attempting to discover the device.

How to Manage Connected *Smart Home* Devices

It is possible to edit a device's name, disable a device, or delete a device.

1.) Go to the menu in the *ALEXA* app.

2.) Tap **Smart Home**.

3.) Select **Devices**.

4.) Select you *smart home* device.

5.) Select **Edit**.

If you'd like to disable all devices associated to a specific skill, rather than delete the devices one at a time, you can simply disable the skill.

Say, "Disable [skill's name] skill."

You can also:

1.) Go to the menu in the *ALEXA* app

2.) Select **Skills**

3.) Select **Your Skills**

4.) Select a skill.

5.) You should now see a *skill* detail page. Select the *Disable* tab.

How to Control Multiple *Smart* Devices Simultaneously

Controlling multiple *Smart Home* devices at once can be accomplished by setting up "Routines." This will enable you to give *ALEXA* multiple commands at once.

To illustrate:

Instead of telling *ALEXA* to turn on a lamp, and then telling *ALEXA* to turn on a fan several moments later, it is possible to simply say, "*ALEXA*, I'm home." *ALEXA* will then turn on the lamp and fan at the same time, combining two commands into one.

Currently, routines are compatible with the following devices:

• *Amazon Echo*

• *Echo Dot*

• *Echo Show*

• *Echo Plus*

Routines are limited to *Smart* home control, news, weather, and traffic.

Create a Routine

1.) Using your mobile device (*Android* or *IOS*), open the *ALEXA* app.

2.) Open *Settings*.

3.) Select **Routines** from the drop-down menu.

4.) Select the + sign at the top-right corner of the screen.

5.) Tap **When this happens,** then select **When you say something**.

If you'd like to set a time and choose which days it should repeat, you can also select **At a scheduled time.** Then select **Done**.

6.) On the *When you say something* page, enter a phrase, such as "*Movie Night,*" or "*Sunrise.*" Then select the **Done** tab.

7.) Now select **Add action**.

8.) You now have a choice to select *Weather, Traffic,* or *News*. If you'd like to power *smart* lights on or off, unlock *smart* doors, etc, select **Smart Home**. Then select **Add**.

9.) Repeat the process until you have added all of the desired actions to the routine.

10.) Under *The device you speak t*o, select which speaker the routine should play audio from.

11.) Select **Create** to complete setting up the routine.

To edit a routine:

1.) Go to *Settings*.

2.) Go to *Routines*.

3.) Select *Edit Routines*.

The routine can also be temporarily disabled in the *Edit Routines* section.

<u>Shades and Colors for *Smart* Light Control</u>

It is possible to use *ALEXA* to change the colors and shades of white for light bulbs, as long as the light bulbs are compatible.

<u>To see if your *smart home* light bulb supports colors, use the *ALEXA* app.</u>

1.) Go to the menu and select **Smart Home**.

2.) Select the compatible light bulb.

3.) Select **Edit**.

4.) Check the **About** section. You should then be able to see all the supported color capabilities.

When it's good to go, simply say, "*ALEXA,* make the [*smart home* device / group name] warmer / cooler to incrementally adjust shades of white.

Or say, "*ALEXA,* set the lights to [color]."

<u>*Smart Light colors can also be changed through the ALEXA app*</u>

1.) Go to the menu and select **Smart Home**.

2.) Select a compatible light bulb.

3.) Select **Set Color**.

<u>Shades of White</u>

• White

• Soft White

• Daylight

• Cool White

• Warm White

<u>Colors</u>

• Gold

• Crimson

• Cyan

• Blue

• Green

• Lavender

• Red

- Salmon

- Yellow

- Violet

- Sky Blue

- Teal

- Turquoise

- Pink

- Orange

- Lime

- Purple

There are many more color options that can be viewed via the *ALEXA* app when you go to **Set Color**.

How to Use *Echo* with *IFTTT*

*I*FTTT stands for *If This Then That*. It's an online service that uses rules (applets) to connect a variety of apps and devices together.

Generally, it helps users do more with their apps and devices.

The *ALEXA* device supports *IFTTT*, and it can trigger the *IFTTT* "rules" you have activated.

To illustrate:

If you ask *ALEXA* to find your phone, *IFTTT* can trigger your phone to ring. Or if you'd like have your *android* phone muted at bedtime, that can be done also.

New applets can be created, or you can choose from applets that already exist from other *IFTTT* users.

1.) If you haven't done so already, go to the *IFTTT* website and sign up.

2.) On the *IFTTT* website, find the *ALEXA* app by typing it into the search bar.

3.) Select the *Connect* tab when the *Amazon ALEXA* page comes up.

4.) Sign in to your *Amazon* account.

After signing in, your *Amazon* account should now be linked to your *IFTTT* account.

You have the option to remove the link between *ALEXA* and *IFTTT* at any time by visiting *Manage Login with Amazon*.

How to Create an Applet

1.) Click *invent your home* at the top-right corner of the screen on your *IFTTT* page.

2.) Select *New Applet*. Then click + *This*. *If This* is the trigger part of the process.

3.) On the *Choose a service* page, type *ALEXA* into the search bar and select it.

4.) You should now be on the *Choose trigger* page. If you'd like to customize the wording, scroll down to the box that reads, *Say a specific phrase*. For example, to turn off the lights, you might want to say, "Power off," instead of "Turn off the lights," so that's what you would type in.

5.) After choosing a phrase to trigger the action, select *Create trigger*.

6.) Click on + *That*. *If That* is the action part of the process. Then choose an action. For example, if you are using *WEMO* light bulbs and you'd like to power them on and off through *ALEXA*, search for *WEMO* in the search bar. In this case, you would then go to *WEMO lighting* and select *Connect*.

Although different apps have different connection methods, most of them are rather similar.

7.) After it's connected, go back and finish selecting an action. Following the previous example, if you have chosen *WEMO* light bulbs, you can now choose what you'd like to happen (Dim the light, Dim a group of lights, etc.) Then select the *Create action* tab.

8.) Select *Finish*.

To power the lights off, you would have to repeat the above steps. This time, on the *Create trigger* page, you can type in an "off" trigger, such as, "Power off."

There are plenty more things you can do, but the process generally remains the same. Scroll through the *IFTTT* lists and find something that interests you. When you are ready to set something up, follow the above steps outlined in this chapter.

ALEXA Command and Request List

Basics

"ALEXA, turn up the volume."

"ALEXA, turn down the volume."

"ALEXA, let's chat."

"ALEXA, stop."

"ALEXA, go to sleep."

"ALEXA, help."

Music

"ALEXA, next song."

"ALEXA, skip song."

"ALEXA, previous song."

"ALEXA, pause in [room name]."

"ALEXA, resume in [room name]."

"ALEXA, play the next track in [room name]."

"ALEXA, louder in [room name]."

"ALEXA, quieter in [room name]."

"ALEXA, set the volume to [volume number or percentage] in [room name]."

"ALEXA, mute [room name]."

"ALEXA, turn it up in [room name]."

"ALEXA, what's playing in [room name]?"

"ALEXA, play music by [artist]."

"ALEXA, what's this song?"

"ALEXA, buy [album name] by [artist's name]."

"ALEXA, play the top songs this week."

"ALEXA, play my [playlist name] playlist."

"ALEXA, shuffle my new music."

"ALEXA, shop for new music by [artist's name]."

"ALEXA, play the [station name] on [music service name]."

"ALEXA, add this song."

"ALEXA, who sings the song [song title]?"

"ALEXA, who is in the band [band's name]?"

"ALEXA, sample songs by [artist]."

To-do lists

"ALEXA, add [item] to my shopping list."

"ALEXA, create a to-do list."

"ALEXA, put [task] on my to-do list."

"ALEXA, I need to [task]."

Shopping on *Amazon*

"ALEXA, add [item] to my cart."

"ALEXA, buy [product]."

"ALEXA, order [item]."

"ALEXA, reorder [item]."

"ALEXA, where's my stuff?"

"ALEXA, track my order."

Smart Home

"ALEXA, discover my *smart home* devices."

"ALEXA, BLUETOOTH."

"ALEXA, connect to my phone."

"ALEXA, is the front / back door locked?"

"ALEXA, lock the front / back door."

"ALEXA, turn on the lights."

"ALEXA, turn on the TV."

"ALEXA, raise the temperature [number] degrees."

"ALEXA, set the temperature to [number]."

"ALEXA, what's the temperature in here?"

"ALEXA, what's the thermostat set to?"

"ALEXA, make the living room [color]."

"ALEXA, turn the desk lamp to [color]."

"ALEXA, turn on the hallway light."

"ALEXA, turn on *Movie Time*."

"ALEXA, dim the living room to [percentage]."

"ALEXA, set the fan to [percentage]."

Weather

"ALEXA, what's the weather in [name of city]."

"ALEXA, what's the temperature?"

"ALEXA, what will the weather be like in [name of city] tomorrow?"

"ALEXA, what's the extended forecast for [name of city]."

"ALEXA, is it going to rain today?"

"ALEXA, will it snow tomorrow?"

"ALEXA, will I need an umbrella today?"

Traffic and Local Information

"ALEXA, how is traffic?"

"ALEXA, what's my commute like?"

"ALEXA, what are the business hours of [venue name]?"

"ALEXA, what [venues] are nearby?"

"ALEXA, what time is the movie, [film name] playing?"

"ALEXA, find the address for [place]."

"ALEXA, is [venue] open?"

"ALEXA, ask *UBER* to request a ride."

News

"ALEXA, what's in the news?"

"ALEXA, give me my flash briefing."

"ALEXA, open [publication name]."

"ALEXA, pause."

"ALEXA, next."

"ALEXA, previous."

Sports

"ALEXA, give me my sports update."

"ALEXA, what was the score of the [name of team] game?"

"ALEXA, did the [team's name] win?"

"ALEXA, when do the [team's name] play next?"

Alarm Clock

"ALEXA, set an alarm for [time]."

"ALEXA, when's my next alarm?"

"ALEXA, snooze."

"ALEXA, set a timer for [time length]."

"ALEXA, set a second timer for [time]."

"ALEXA, cancel my alarm for [time]."

"ALEXA, what time is it?"

"ALEXA, cancel all alarms."

"ALEXA, set a repeating alarm for [time] [days]."

Calendar

"ALEXA, what's the date?"

"ALEXA, add an event to my calendar."

"ALEXA, add a [time and event] to my calendar.

"ALEXA, what's on my calendar today?"

"ALEXA, what's my next appointment?"

Knowledge

"ALEXA, how tall is [name of mountain]?"

"ALEXA, how deep is [name of ocean]?"

"ALEXA, what's the capital of [place]?"

"ALEXA, what's the population of [place]?"

"ALEXA, who wrote [name of book]?"

"ALEXA, what's the definition of [word]?"

"ALEXA, how do you spell [word]?"

"ALEXA, what's [number] times [number]?"

"ALEXA, [number] factorial?"

AUDIOBOOKS

"ALEXA, play [book title] on *Audible*."

"ALEXA, pause."

"ALEXA, resume."

"ALEXA, next chapter."

"ALEXA, previous chapter."

"ALEXA, go to last chapter."

"ALEXA, go to [chapter number]."

Just for Fun

"ALEXA, tell me a joke."

"ALEXA, sing a song."

"ALEXA, tell me a story."

"ALEXA, play a game."

Troubleshooting

Many problems can be solved by simply restarting (unplugging the device and plugging it back in) the *Echo*.

But when restarting *Echo* does not correct the issue, there are also some other things that can be tried.

ALEXA App Doesn't Seem to Work

- Confirm your device meets the requirements
- Restart your phone
- If you are using a web browser, close the web browser, then reopen it
- Close the app, then reopen it
- Uninstall the app, then reinstall it

Problems with ALEXA Skills

- Disable the skill, then re-enable it

Streaming Issues

- Verify your internet connection is at least 512 KBPS (0.51 MBPS)
- Turn off other devices that might be absorbing the bandwidth
- Move the *ALEXA* device closer to the router and modem
- Move the *ALEXA* device away from microwaves and other possible sources of interference
- Restart the router and modem

ALEXA Isn't Understanding The Words You're Saying

- If there is background noise, wait for it to clear before speaking to *ALEXA*
- If the *ALEXA* device is on the ground, try moving it to a higher location
- Be more specific
- Use voice training. See the *How to Teach ALEXA Your Voice* chapter.

BLUETOOTH Issues

- Verify your *BLUETOOTH* device uses a supported profile (A2DP SNK, A2DP SRC, AVRCP)

• Move the *BLUETOOTH* device away from microwaves and other possible sources of interference

• Make sure the *BLUETOOTH* device is close enough to the *ALEXA* device when you pair it

• Check the batteries and replace them if necessary

• Clear all the *BLUETOOTH* devices, then restart *ALEXA* and the *BLUETOOTH* device

To clear the *BLUETOOTH* devices with *Echo:*

1.) Open the *ALEXA* app.

2.) Go to the menu and select *Settings*.

3.) Select your device.

4.) Select **BLUETOOTH**.

5.) Select a device from the list.

6.) Select **Forget**.

7.) Repeat the process for all other *BLUETOOTH* devices.

Don't forget to restart the *ALEXA* device and the *BLUETOOTH* device when you are finished clearing the devices.

Smart Home Camera Issues

• Verify the camera is compatible with *ALEXA*

• Verify the *smart home* camera is powered on

• Check the battery or power supply

• Verify you have completed the setup process through the camera manufacturer's website or companion app

• Check for available software updates and install them if necessary

• Check your camera's network settings

• Restart your devices

How to Reset The *Echo 2nd Generation* Device

I f the *Echo* becomes unresponsive, even after you have restarted the device, it is possible to reset it.

Note: Resetting the *Echo* device will require you to register it to an *Amazon* account again, and you will need to reenter any device settings that were previously in effect.

1.) Press and hold the **Microphone off** and **Volume down** buttons simultaneously for approximately 20 seconds until the light ring changes to orange. After changing to orange, the light ring will then turn blue.

2.) Standby while the light ring powers off and back on again. The light ring will shift back to orange, indicating the device is now in setup mode.

3.) Open the *ALEXA* app to connect the *Echo* to a WI-FI network.

4.) Register the device to your *Amazon* account.

How to Use The *ALEXA* Voice Remote (Sold Separately)

T he *ALEXA* Voice Remote is a voice-enabled, battery-powered remote that features a directional track pad, a microphone, and a *talk* button. It allows you to rapidly control audio playback on the *Echo* device.

It must be purchased separately.

Before it is able to be used, it must be paired with the *Echo* device.

Pairing *ALEXA* Voice Remote with *Echo*

1.) Open the latch on the battery door of the remote. The latch can be opened by pulling down on the battery door, then pulling the door away from the remote.

2.) Insert two AAA batteries, then close the latch.

3.) Open the *ALEXA* app.

4.) Go to the menu, then select *Settings*.

5.) Press and hold the **Play/Pause** button on the remote for approximately five seconds, then release it.

The *Echo* device should now search for the remote and connect it within forty seconds or so.

ALEXA will say, "Your remote has been paired," when the remote is discovered by the device.

More From Emery H. Maxwell

Fire HD 10 Tablet Manual, available at all *Amazon* stores, including <u>U.S.</u>